The Eiffel Tower

A Monumental Feat of Engineering

Authored by
Zahid Ameer

Published by

Goodword eBooks

Copyright © 2024 Zahid Ameer

All rights reserved.

ISBN: 9798346013648

Eiffel Tower: A Monumental Feat of Engineering

DEDICATION

"I dedicate this book to my beloved parents, whose wisdom I hold in the highest regard. Their every word of guidance has been a beacon of light, illuminating the path of my life and shaping the essence of who I am."

Eiffel Tower: A Monumental Feat of Engineering

Eiffel Tower: A Monumental Feat of Engineering

Contents:

Contents:

Introduction

Chapter 1: The Visionary Behind the Tower

Chapter 2: Raw Materials Used in the Construction

Chapter 3: Manpower Behind the Monument

Chapter 4: Time and Precision: The Construction Timeline

Chapter 5: The Eiffel Tower's Legacy

Conclusion

Bibliography

Acknowledgments

Disclaimer

About me

Eiffel Tower: A Monumental Feat of Engineering

Eiffel Tower: A Monumental Feat of Engineering

Introduction

The Eiffel Tower, an icon of modern engineering and a symbol of France, is one of the most recognizable landmarks on the planet. Since its completion in 1889, this towering structure has captured the imagination of millions of people around the world. Standing at an impressive height of **324 meters (1,063 feet)**, the Eiffel Tower dominates the Paris skyline, embodying the spirit of innovation and the industrial advancements of the late 19th century. What was once criticized by some as an "eyesore" has now become a beloved cultural and architectural masterpiece, celebrated for its beauty, complexity, and technological significance.

The story behind the Eiffel Tower is not just one of art and architecture, but also of sheer engineering brilliance. It is a story of overcoming challenges—both technical and social—and showcasing human ingenuity at a time when industrialization was transforming societies across Europe and beyond. Conceived as a temporary structure for the **1889 Exposition Universelle** (World's Fair), held in Paris to celebrate the 100th anniversary of the French Revolution, the Eiffel Tower was designed to showcase the capabilities of modern iron construction. It was to be the tallest man-made structure in the world, a title it would

Eiffel Tower: A Monumental Feat of Engineering

hold until the completion of the **Chrysler Building** in New York in 1930.

However, the significance of the Eiffel Tower goes far beyond its height or its function as a World's Fair centerpiece. It is a testament to the ambitious spirit of its creator, **Gustave Eiffel**, a visionary civil engineer who pushed the boundaries of what was thought possible in the field of construction. His design broke new ground, proving that structures could reach unprecedented heights while maintaining both stability and aesthetic appeal. More than that, the Eiffel Tower stands as a symbol of the age of iron and steel—materials that revolutionized the way buildings, bridges, and monuments were constructed during the 19th century.

Despite its current status as an architectural marvel, the Eiffel Tower's journey to acceptance was not without controversy. In the years leading up to its construction, the proposed design was met with fierce opposition from some of Paris' most prominent artists and intellectuals, who feared that the massive iron structure would ruin the city's classical aesthetic. A famous protest letter, known as the "Protests by the Artists Against the Tower of Mr. Eiffel," published in **Le Temps** newspaper in 1887, described the tower as a "gigantic black factory chimney" and accused Eiffel of sacrificing artistic beauty for engineering prowess. Yet, despite these early objections, the tower was

Eiffel Tower: A Monumental Feat of Engineering

built and would go on to win over its critics, becoming an enduring symbol of both Paris and modernity itself.

The construction of the Eiffel Tower was an **extraordinary feat** in every sense of the word. Not only did it require unprecedented amounts of raw materials, but it also involved meticulous planning and execution by a dedicated team of engineers, ironworkers, and laborers. The tower was assembled using over **7,300 tons of wrought iron**, held together by **2.5 million rivets**, and supported by a sophisticated system of foundations that anchored the structure deep into the ground. The sheer scale of the project was unlike anything the world had seen before, and the logistical challenges were immense. The assembly process, carried out in the heart of Paris, required coordination and precision, as workers maneuvered massive iron girders into place, often at dizzying heights.

In addition to the raw materials used, the manpower involved in constructing the Eiffel Tower was vast. A team of **around 300 workers**—including skilled ironworkers, fitters, carpenters, and support staff—labored for over two years to complete the structure. The work was dangerous, with the constant risk of falls or accidents, but thanks to the implementation of safety measures by Eiffel and his team, the construction saw remarkably few fatalities for a project of such scale. The workers' dedication and craftsmanship

Eiffel Tower: A Monumental Feat of Engineering played an essential role in turning Eiffel's ambitious vision into a reality.

The Eiffel Tower's construction timeline is equally impressive. Despite the immense scale of the project, the tower was completed in **just over two years**, with work beginning in January 1887 and finishing in March 1889. This rapid pace was achieved through a combination of meticulous pre-fabrication, on-site assembly, and innovative engineering techniques. The tower was designed to be modular, with individual sections of iron prepared in a factory and then transported to the construction site, where they were riveted together. This method allowed the workers to assemble the tower piece by piece, steadily raising the structure to its final height.

As we delve deeper into the specifics of the Eiffel Tower's construction in the following chapters, we will explore in detail the raw materials that went into its creation, the labor force that made it possible, and the precise timeline that brought this remarkable structure to life. Each element of the project—from the sourcing of the iron to the placement of the final rivet—was carefully planned and executed, reflecting the unparalleled ambition of the engineers and workers involved.

By examining the Eiffel Tower through the lens of its construction, we gain not only an appreciation for the

Eiffel Tower: A Monumental Feat of Engineering

technical challenges it overcame but also a deeper understanding of its enduring significance. The Eiffel Tower is more than just a monument or a tourist attraction; it is a **masterpiece of engineering**, an enduring symbol of the human capacity to innovate, and a lasting reminder of an era when industrial progress was reshaping the world.

As you read on, this book will guide you through the step-by-step process of how this iconic structure was built, from the foundation to the tip of its spire. We will explore the extraordinary quantities of raw materials that were consumed, the incredible manpower that made it possible, and the timeline of events that culminated in one of the greatest architectural achievements of the 19th century. This is the story of how the Eiffel Tower—one of the world's most admired and enduring landmarks—came into being, and the lasting impact it continues to have on architecture, engineering, and culture to this day.

Eiffel Tower: A Monumental Feat of Engineering

Chapter 1: The Visionary Behind the Tower

The construction of the Eiffel Tower, an architectural marvel of the late 19th century, would not have been possible without the visionaries who brought it to life. At the heart of this project was **Gustave Eiffel**, a civil engineer whose name is forever tied to the tower, but behind him were also brilliant minds like **Maurice Koechlin** and **Émile Nouguier**, who played critical roles in its design and conception. Together, they crafted a structure that would not only symbolize France but also stand as a testament to human ingenuity. However, their journey was far from easy, and the tower's path from idea to reality was filled with challenges, doubts, and fierce opposition.

Gustave Eiffel: The Mastermind

Gustave Eiffel was already an accomplished engineer by the time the idea of the tower was born. He had gained a reputation for his work on ambitious structural projects, including bridges, viaducts, and the internal framework for the Statue of Liberty. Born in **Dijon, France, in 1832**, Eiffel developed an early interest in metal construction, which led him to study at the prestigious **École Centrale des Arts et Manufactures** in Paris. His engineering firm,

Eiffel Tower: A Monumental Feat of Engineering

Compagnie des Établissements Eiffel, had already undertaken major projects worldwide, making him a well-known figure in the field of civil engineering.

Eiffel's expertise in using iron in construction, particularly his mastery of iron lattice frameworks, made him the perfect candidate to undertake an audacious new project. His career had been a journey of experimentation and innovation, from constructing the **Garabit Viaduct** in central France, one of the highest bridges in the world at the time, to helping create the **Statue of Liberty's** internal iron structure. These projects demonstrated his vision of combining practicality with grandeur, which would be fully realized with the Eiffel Tower.

Maurice Koechlin and Émile Nouguier: The Unsung Engineers

While Gustave Eiffel is often celebrated as the architect of the Eiffel Tower, the initial concept for the structure came from two of his chief engineers—**Maurice Koechlin** and **Émile Nouguier**. Both men worked for Eiffel's company and were responsible for designing many of the firm's large-scale projects. In 1884, they began brainstorming ideas for a monument that could be constructed as a showcase for French engineering at the upcoming **1889 Exposition Universelle**, or **World's Fair**, which would mark the **100th anniversary of the French Revolution**.

Eiffel Tower: A Monumental Feat of Engineering

Koechlin and Nouguier envisioned a tall iron structure that would demonstrate the potential of iron as a building material for monumental architecture. Their preliminary sketches showed a skeletal framework composed of iron beams and arches, rising in a tapering fashion to a height never before achieved. Though the initial idea was theirs, they knew they needed the support of Eiffel, the company's figurehead and influential name in engineering, to bring the project to life.

When Koechlin presented the first draft of the design to Eiffel in June 1884, Eiffel was intrigued but cautious. He suggested several changes to refine the concept, and with the help of **architect Stephen Sauvestre**, the design was further developed into the form we recognize today. Sauvestre added decorative arches at the base, a glass pavilion on the first level, and other embellishments to make the tower more visually appealing. By December of that year, Eiffel had fully embraced the project, recognizing its potential as both a technical and artistic achievement.

The 1889 World's Fair: A Stage for French Innovation

The **Exposition Universelle of 1889** was to be one of the most prestigious world's fairs ever held. It aimed to celebrate the centennial of the **French Revolution** and showcase France's cultural and industrial accomplishments

Eiffel Tower: A Monumental Feat of Engineering

to a global audience. Paris had already established itself as a center of art, fashion, and intellectual thought, but in the rapidly industrializing world of the late 19th century, the exposition offered a chance to display French prowess in the realms of science and engineering. The fair needed a centerpiece, something that would capture the imagination of visitors and symbolize France's leap into the modern age.

Eiffel, now firmly behind the project, saw an opportunity in the tower. Not only would it be a technical triumph, standing taller than any man-made structure at the time, but it would also serve as a symbol of France's industrial leadership. He secured the rights to build the tower after signing a contract with the French government, agreeing to fund much of the construction himself in exchange for control over the tower's profits for the first 20 years. With the plans finalized and the contract signed in **1886**, construction was set to begin.

Public Skepticism and Opposition

Despite Eiffel's enthusiasm and the technical brilliance of Koechlin and Nouguier's design, the project faced immediate and fierce opposition from many quarters. **Parisians**—particularly the artistic and intellectual elite—were deeply divided over the idea of erecting a colossal iron tower in the heart of their beloved city. At the time,

Eiffel Tower: A Monumental Feat of Engineering

Paris was renowned for its elegant, classical architecture, and many feared that the tower would clash with the city's aesthetic. Some even went so far as to call the design **"monstrous"** and **"barbaric"**, believing it would ruin the city's skyline.

One of the most vocal groups of opponents included artists and writers, who saw the tower as an affront to Parisian culture. In **1887**, a group of prominent artists, including **Guy de Maupassant** and **Charles Gounod**, penned a protest letter, famously known as **"The Protest Against the Tower of Mr. Eiffel"**. In this letter, they decried the tower as a **"gigantic black factory chimney"**, stating that it would disfigure the city and undermine its historic charm. The letter, published in the newspaper **Le Temps**, read in part:

"We, writers, painters, sculptors, architects, lovers of the beauty of Paris, solemnly protest against the erection, in the heart of our capital, of the useless and monstrous Eiffel Tower... We see it as the dishonor of Paris."

Eiffel, however, was undeterred by the criticism. He defended the project vigorously, often pointing out the scientific and technological benefits that the tower would offer. He emphasized that the tower would serve as a demonstration of the advances in iron construction, and as

Eiffel Tower: A Monumental Feat of Engineering a symbol of France's modernity and forward-thinking spirit. Eiffel once famously remarked:

"Do you think it is because of the artistic value of the pyramids that they have so strongly held the attention of men? What is admirable in the great pyramids of Egypt is the strength of the conception, more than the height of the monuments themselves."

To Eiffel, the tower was more than just an architectural structure—it was a statement of human achievement, one that would transcend artistic conventions and pave the way for future developments in architecture and engineering.

The Approval and the Beginning of Construction

After much debate, the French government gave the project the green light in 1887, and construction officially began on **January 28, 1887**. Eiffel had convinced both the public and the authorities that the tower would not only be a centerpiece for the upcoming World's Fair but also an enduring symbol of French ingenuity. Eiffel's unwavering confidence in the project was finally rewarded, and his company was tasked with building the structure in an ambitious timeline—just over two years.

Gustave Eiffel's Role as Leader

Eiffel Tower: A Monumental Feat of Engineering

Eiffel was much more than just a financier or figurehead for the tower. He played an active role in overseeing the construction, ensuring that every detail was executed to the highest standards. His hands-on approach was evident throughout the process, from designing the foundation work to ensuring the iron pieces were manufactured and assembled with precision. Despite the immense scale of the project, Eiffel was meticulous about safety, implementing strict measures to protect his workers and minimize accidents—an unusual practice for large-scale projects at the time.

Eiffel's ability to manage both the technical challenges of the project and the public relations aspect of the tower was critical to its success. He weathered the storms of criticism and skepticism with determination, often appearing on-site to personally oversee the work and inspire his team.

A Lasting Legacy

The Eiffel Tower was completed in **March 1889**, just in time for the **World's Fair**. When it was unveiled, the world was stunned. The tower, which stood at an unprecedented height of **300 meters** (984 feet) at the time, was the tallest man-made structure in the world, a record it held until the completion of the **Chrysler Building** in New York City in 1930. Despite the initial controversy, the tower quickly became a symbol of progress and modernity,

Eiffel Tower: A Monumental Feat of Engineering capturing the imaginations of visitors from around the globe.

Over time, the Eiffel Tower's critics were silenced by its success and its status as an iconic Parisian landmark. Gustave Eiffel's vision had been vindicated, and what was once seen as an eyesore became a beloved and enduring symbol of France. To this day, the Eiffel Tower stands as a reminder of the power of bold ideas, perseverance, and the forward march of engineering innovation.

Gustave Eiffel himself continued to work on other projects, including experimenting with aerodynamics and contributing to the early development of radio transmission technology. However, the Eiffel Tower remains his crowning achievement—a monument that bears his name and continues to inspire awe and admiration more than 130 years later.

Chapter 2: Raw Materials Used in the Construction

The construction of the **Eiffel Tower** was a monumental engineering endeavor, requiring a carefully chosen array of raw materials to ensure that the structure would not only be impressive but also strong, durable, and resistant to the elements. Each material used in the construction played a crucial role in the tower's design, stability, and longevity. At the core of the project was **wrought iron**, a material selected for its unique properties, while other elements such as rivets, concrete, and paint were essential in holding the structure together and preserving it over time. This chapter delves deeply into the raw materials that made the Eiffel Tower possible.

Wrought Iron: The Backbone of the Eiffel Tower

The Eiffel Tower was primarily constructed using **wrought iron**, which was the most practical and readily available material at the time. Steel, which would later become the dominant material in high-rise construction, was not yet in widespread use, and wrought iron offered a cost-effective, durable alternative that could be molded into the intricate lattice design that Gustave Eiffel envisioned.

Eiffel Tower: A Monumental Feat of Engineering

- **Quantity**: Approximately **7,300 tons** of wrought iron were required to build the Eiffel Tower. The sheer amount of iron used highlights the immense scale of the project. For context, this is roughly the weight of 1,500 large elephants or 1,000 cars. The iron was used to create over **18,000 individual pieces**, all of which had to be precisely fabricated and then assembled on-site.
- **Source**: The iron was sourced from the **Pompey factory** in Lorraine, France. This factory was known for producing high-quality iron, and it played a critical role in supplying the materials for one of the largest construction projects of the time. The decision to use wrought iron was not just based on its strength but also its availability in large quantities from domestic sources. This helped to reduce costs and logistical complexities during the construction process.
- **Purpose**: Wrought iron was chosen for its **ductility**, **malleability**, and **strength**. The Eiffel Tower's innovative lattice structure required a material that could be shaped into complex forms without sacrificing durability. Wrought iron fulfilled this requirement perfectly, allowing the engineers to create a lightweight yet incredibly strong framework. The lattice design, with its open, airy structure, helped to reduce the overall weight of the

Eiffel Tower: A Monumental Feat of Engineering

tower while maintaining its stability, enabling it to rise to unprecedented heights. The iron pieces were pre-fabricated into precise shapes, making assembly more efficient.

The **iron framework** is the defining characteristic of the Eiffel Tower's appearance. The lattice design not only gives the tower its iconic look but also distributes weight evenly, ensuring that the structure could withstand strong winds, changes in temperature, and other environmental forces. The result was a structure that was light enough to be built to a height of over 300 meters while remaining strong and resilient. This was an engineering feat that had never been attempted on such a scale before.

Rivets and Bolts: The Hidden Heroes of the Eiffel Tower

While the wrought iron provided the basic structure, it was the **rivets and bolts** that held everything together. Without these fasteners, the tower's lattice framework would not have been possible.

- **Quantity**: Around **2.5 million rivets** were used to join the iron pieces together. Each rivet was essential in securing the individual iron sections, ensuring that the structure was both strong and flexible. The use of such a vast number of rivets was necessary to

Eiffel Tower: A Monumental Feat of Engineering

distribute the load across the entire structure, preventing weak points from developing.

- **Significance**: Riveting was a painstaking and labor-intensive process. The rivets were heated until red-hot in portable forges on the construction site. Once heated, they were inserted into pre-drilled holes in the iron plates, where they were hammered into shape, creating a strong and permanent bond. This technique was not only critical for ensuring the structural integrity of the tower but also allowed for some flexibility. The Eiffel Tower was designed to sway slightly in the wind, and the riveted joints allowed for this movement without compromising the overall stability of the structure.

The use of rivets instead of bolts or welds was a deliberate choice. Rivets, when installed correctly, create a joint that is both tight and flexible, allowing for slight movement without breaking. This was particularly important for a structure as tall and exposed to the elements as the Eiffel Tower. The ability to withstand high winds, temperature changes, and even minor ground movements was essential to the tower's longevity.

Concrete Foundations: The Anchor of the Iron Giant

The Eiffel Tower's four massive legs required **deep foundations** to anchor the structure securely into the

Eiffel Tower: A Monumental Feat of Engineering

ground. This was especially important given the immense weight of the tower and the variable nature of the soil beneath it.

- **Materials Used**: The foundations were made primarily of **concrete**, which was poured into trenches dug beneath each of the tower's four legs. The amount of concrete required was substantial, as the foundations needed to be strong enough to support the weight of the entire structure while remaining stable even under changing environmental conditions. Each foundation was carefully designed based on the specific characteristics of the soil in that area.
- **Soil Considerations**: The soil beneath the Eiffel Tower varied in composition, which posed a challenge for the engineers. Two of the legs were positioned on solid ground, which required only relatively shallow foundations. However, the other two legs were located on softer, more unstable soil, necessitating much deeper foundations to ensure stability. The engineers had to take great care in designing these foundations to ensure that the tower would not shift or settle unevenly over time.

The **concrete foundations** ensured that the Eiffel Tower was rooted firmly in place, capable of supporting the tremendous weight of the iron superstructure above. These

Eiffel Tower: A Monumental Feat of Engineering

foundations remain one of the critical, yet often overlooked, aspects of the tower's construction. Without them, the tower could not have achieved its height or stability.

Paint: Protecting the Iron from Time and the Elements

One of the most important, yet least celebrated, aspects of the Eiffel Tower's construction was the extensive use of **paint** to protect the iron from corrosion. Over time, unprotected iron is highly susceptible to rust, which could have compromised the tower's structural integrity.

- **Quantity**: The original construction of the Eiffel Tower required over **60 tons of paint** to cover its entire surface. Given the size of the structure and the complexity of its lattice design, painting the Eiffel Tower was a massive undertaking in itself. Every inch of the iron framework had to be carefully coated to ensure it was protected from the elements.
- **Purpose**: The primary reason for painting the Eiffel Tower was to protect the iron from **rust and corrosion**. Paris's humid climate, along with pollution, could have caused the iron to deteriorate rapidly if it had been left unprotected. By applying multiple layers of paint, the engineers were able to create a barrier that shielded the iron from moisture and the corrosive effects of the environment.

Eiffel Tower: A Monumental Feat of Engineering

- **Repainting the Tower**: The Eiffel Tower has been repainted about once every seven years since its completion. This regular maintenance is essential to preserve the structure and ensure its longevity. Over the years, the color of the paint has changed several times. Initially, the tower was painted a reddish-brown color, but it has since been coated in various shades of yellow, brown, and gray. Today, the Eiffel Tower is painted in a color known as "Eiffel Tower Brown," a specially mixed hue that is designed to blend harmoniously with the Paris skyline.

The process of painting the Eiffel Tower is no small feat. It requires a team of workers who must scale the structure and apply the paint by hand, ensuring that every rivet, beam, and joint is adequately covered. The paint not only serves a protective function but also enhances the aesthetic appeal of the tower, ensuring that it remains a striking and well-maintained landmark for future generations.

Conclusion

The Eiffel Tower's construction relied on a delicate balance of materials, each chosen for its unique properties and role in the overall design. **Wrought iron** provided the strength and flexibility needed to build a structure of such unprecedented height, while **rivets** and **bolts** ensured that the iron pieces were securely held together. The **concrete**

Eiffel Tower: A Monumental Feat of Engineering

foundations anchored the tower to the ground, and **paint** protected it from the ravages of time and weather. Together, these materials formed the backbone of the Eiffel Tower, allowing it to stand tall and proud as one of the most iconic structures in the world.

This careful selection and use of materials are part of what makes the Eiffel Tower not only an engineering marvel but also a symbol of human innovation and perseverance. The raw materials may be simple—iron, concrete, and paint—but in the hands of skilled engineers and laborers, they were transformed into a structure that has captivated the world for over a century.

Eiffel Tower: A Monumental Feat of Engineering

Chapter 3: Manpower Behind the Monument

Building the Eiffel Tower was a formidable challenge that required the concerted efforts of a dedicated team. The vision and design behind this iron marvel were revolutionary for its time, but the actual construction required skilled labor, coordinated logistics, and rigorous safety measures. From the highly skilled ironworkers assembling its intricate lattice structure to the engineers and architects ensuring every detail met exact specifications, the manpower involved was nothing short of remarkable. This chapter delves deep into the human aspect of constructing one of the world's greatest architectural triumphs.

Ironworkers: The Backbone of the Construction

At the heart of the Eiffel Tower's assembly were the **ironworkers**, a group of around **300 skilled laborers** whose expertise lay in working with iron. Their tasks were critical to transforming the raw materials—thousands of tons of wrought iron—into the iconic structure that still stands today. These workers were responsible for piecing together the pre-fabricated iron elements that had been

Eiffel Tower: A Monumental Feat of Engineering prepared off-site, ensuring that each piece fit into place perfectly.

The Number and Roles of Workers

The ironworkers were divided into specific roles:

- **Riveters**: These workers were responsible for fastening the iron elements together using **red-hot rivets**, which were essential in holding the structure securely. Riveting was no small task. A team of four was typically involved in placing each rivet: one person would heat the rivet in a portable forge, another would catch it as it was tossed, and the other two would hammer it into place to secure the iron joints. Given that over **2.5 million rivets** were used in the construction, the riveters played a vital role in the success of the project.
- **Fitters**: Fitters were tasked with aligning and adjusting the iron beams, ensuring that every piece of the latticework fit together perfectly. The precision required was immense, as any slight misalignment could cause the structure to lose its stability. Fitters often had to work at great heights, securing beams while perched precariously on scaffolding or the tower's incomplete frame.
- **Carpenters**: Although the tower was constructed primarily of iron, the carpenters' role was

Eiffel Tower: A Monumental Feat of Engineering

indispensable. They built the **scaffolding** and wooden frameworks that supported the workers as they assembled the iron structure. The carpenters' contributions were especially critical in the early phases of construction, when the tower's foundation and lower levels required support for the heavy materials and workers.

The Challenges of Assembling Iron

The Eiffel Tower was constructed from **18,038 individual iron parts**, all of which were prefabricated and pre-measured off-site. This allowed for rapid assembly on-site, but it also meant that any errors in fitting the parts together could lead to delays or structural issues. The ironworkers had to ensure that each piece fit perfectly into the overall design, which required an extraordinary level of precision and skill. This process became more complex as the tower rose higher, with workers having to hoist heavy iron beams to ever-increasing heights using cranes and pulleys.

Work at Dizzying Heights

As the Eiffel Tower rose higher, the challenges faced by ironworkers multiplied. Working at great heights was both dangerous and exhausting. With each ascending level, the workers had to balance the risks of working hundreds of feet above the ground while maintaining focus on their

Eiffel Tower: A Monumental Feat of Engineering

precise and intricate tasks. By the time the workers reached the **second platform** (about 115 meters high), the task had become even more treacherous.

Despite the inherent dangers, the Eiffel Tower was completed without a significant number of fatal accidents—a testament to the skill and discipline of the workers and the safety measures put in place. Gustave Eiffel was known for caring about his workers' well-being, and he provided safety equipment such as **guardrails** and **safety screens**, which were innovative at the time. This was highly unusual in the 19th century, as worker safety was not typically prioritized in construction projects.

Support Workers: The Logistics of Construction

Behind the scenes, a vast network of **support workers** was crucial to the success of the project. While the ironworkers assembled the structure, these laborers were responsible for various logistical tasks that kept the construction moving smoothly. Their contributions, though less visible, were equally essential.

Material Transport

One of the biggest challenges was transporting the **7,300 tons of wrought iron** and other materials to the construction site. The iron pieces were fabricated in a factory in Pompey, Lorraine, and transported to Paris by

Eiffel Tower: A Monumental Feat of Engineering

train. From the train station, they were moved to the construction site by horse-drawn carts and other means of transportation available at the time. The logistical team handled this complex supply chain, ensuring that the right materials arrived on time, in the right quantities, and in the correct sequence.

Concrete Foundations

Another group of workers was responsible for preparing the **concrete foundations** that supported the tower's massive weight. The foundations needed to be particularly robust due to the soft soil in parts of the site. Workers dug deep into the ground to anchor the legs of the tower, filling these holes with large amounts of concrete. For the legs resting on more stable ground, the foundations were shallower but still required precision in their placement to ensure the tower's balance.

Scaffolding and Temporary Structures

The construction of scaffolding was a key aspect of the support work. The tower's scaffolding, which supported workers as they built the iron frame, was a massive undertaking in itself. Wooden frameworks and platforms had to be constantly adjusted and rebuilt as the tower rose, giving ironworkers access to the higher sections.
Carpenters and **laborers** worked tirelessly to build these

Eiffel Tower: A Monumental Feat of Engineering

temporary structures, ensuring the ironworkers could safely perform their tasks at dizzying heights.

Safety Measures: A Forward-Thinking Approach

Construction during the late 19th century was often dangerous, and worker fatalities were common on large projects. However, Gustave Eiffel's concern for his workers set him apart from many of his contemporaries. Eiffel implemented **groundbreaking safety measures**, which were highly unusual for the time.

Guardrails and Screens

One of the key safety innovations Eiffel introduced was the use of **guardrails** and **safety screens** on scaffolding. These helped prevent falls, which were the most significant risk to workers, especially those working at the highest levels of the tower. The presence of these safety measures was highly appreciated by the workforce and contributed to the relatively low number of accidents during the tower's construction.

Harnesses and Ropes

In addition to guardrails, workers were provided with **harnesses and ropes** to help secure themselves while working at great heights. This early form of fall protection was quite advanced for its time and was another example of Eiffel's commitment to worker safety. As a result of these measures, only one worker died during the two years of construction—a remarkably low number considering the scale and difficulty of the project.

Engineers and Architects: The Minds Behind the Monument

While the ironworkers and support staff handled the physical labor, the tower's engineers and architects were responsible for making sure the structure met all necessary technical requirements. They oversaw the project's design, made adjustments as needed, and ensured that the tower would be structurally sound.

Gustave Eiffel: The Mastermind

At the center of the engineering team was **Gustave Eiffel**, the project's visionary leader. Eiffel's background in civil engineering and his experience working on large-scale projects made him the perfect person to oversee the construction of such an ambitious structure. His personal involvement in the day-to-day progress of the tower was extensive. He would often visit the site, checking measurements and ensuring that the ironworkers were following the plans to the letter.

Eiffel's attention to detail was legendary. He personally oversaw the alignment of the tower's legs and ensured that

Eiffel Tower: A Monumental Feat of Engineering

the curved sections of the lattice framework met his exacting standards. The tower's ability to withstand the forces of wind and weather—something that concerned critics at the time—was the result of careful calculations and testing. Eiffel's commitment to safety, both in terms of the structural integrity of the tower and the well-being of his workers, was central to the project's success.

Maurice Koechlin and Émile Nouguier: The Engineering Geniuses

While Eiffel was the public face of the project, two other engineers played crucial roles in its design: **Maurice Koechlin** and **Émile Nouguier**. Koechlin and Nouguier were responsible for much of the technical planning that went into designing the lattice structure of the tower. Their calculations allowed the tower to be both lightweight and strong, with a frame that could flex slightly in the wind without compromising its stability.

Koechlin and Nouguier's use of **mathematical precision** was key to the success of the project. They meticulously calculated the loads that the tower would need to support, the forces of wind resistance, and the strength of the riveted joints. These calculations enabled the team to create a structure that was both elegant and durable—capable of standing for centuries.

Eiffel Tower: A Monumental Feat of Engineering

Conclusion: A Harmonious Collaboration of Skills

The construction of the Eiffel Tower was a massive undertaking that required the combined efforts of hundreds of skilled workers, support staff, and visionary engineers. The ironworkers, fitters, riveters, and carpenters all played critical roles in bringing Gustave Eiffel's vision to life, while the logistical support ensured the smooth delivery of materials and the preparation of solid foundations. Above all, it was the collaboration between Eiffel and his team of engineers, particularly Maurice Koechlin and Émile Nouguier, that allowed the Eiffel Tower to become a reality.

Despite the numerous challenges—working at great heights, handling heavy iron pieces, and ensuring the structural integrity of a never-before-seen design—the Eiffel Tower was completed in just over two years. Its construction stands as a testament to the incredible skill, ingenuity, and collaboration of the manpower behind the monument.

Eiffel Tower: A Monumental Feat of Engineering

Chapter 4: Time and Precision: The Construction Timeline

The construction of the Eiffel Tower is a fascinating example of engineering brilliance and precise planning. Despite the magnitude of the project, the tower was completed in a surprisingly short period of just over two years. Given the challenges posed by the size and height of the structure, the Eiffel Tower's construction stands as a triumph of 19th-century technology and organization. This chapter takes an in-depth look at the key stages of the tower's construction, from the initial preparations to the final piece being set in place.

Initial Preparations

Start Date: January 28, 1887

Before the actual construction of the Eiffel Tower could commence, there was a considerable amount of preparatory work that needed to be done. These preliminary stages were critical to ensuring that the tower's foundations would support the immense weight and height of the structure.

Surveying the Site

Eiffel Tower: A Monumental Feat of Engineering

The chosen site for the Eiffel Tower was on the **Champ de Mars**, a large public green space located near the Seine River in Paris. This area had been selected because it was both spacious and central, offering ample room for such a massive project. However, building on this site presented its own challenges, especially with regard to the soil conditions.

To begin with, **surveyors** had to meticulously analyze the ground to determine the best way to anchor the tower. Given the sheer size of the structure, it was imperative that the foundations be stable and able to bear the weight of the tower, which would ultimately reach **10,100 tons** when completed.

Laying the Foundations

The Eiffel Tower rests on four massive legs, and the stability of these legs was essential to the overall success of the project. Each of the four legs required large foundations to support the structure's weight, but the soil beneath them varied in composition.

- **Two of the legs** on the Champ de Mars side were situated on solid ground, which allowed for relatively straightforward construction of their foundations.
- However, the other **two legs**, closer to the Seine River, were located on softer, more unstable soil.

Eiffel Tower: A Monumental Feat of Engineering

This presented a significant challenge for the engineers. To address this issue, they had to dig deeper into the ground and use **compressed air caissons** to stabilize the foundations and prevent shifting over time.

The foundations were made from **massive concrete blocks**, each capable of bearing the immense pressure exerted by the tower's weight. The process of digging, setting up caissons, and pouring concrete took about **five months**. By the time the foundations were completed in **June 1887**, the team had laid the groundwork for what was to come.

Prefabrication of Iron Components

While the foundations were being prepared, another crucial aspect of the project was already underway: the **prefabrication of the iron parts**. One of the most remarkable aspects of the Eiffel Tower's construction is that much of it was prefabricated offsite. The iron parts were forged at **Maurice Koechlin's workshop** in **Lorraine**, and then shipped to the site in Paris.

This prefabrication process was highly innovative for the time, as it allowed the builders to save considerable time during the assembly phase. By the time the foundations were ready, thousands of prefabricated parts were waiting

Eiffel Tower: A Monumental Feat of Engineering

to be assembled on-site, ensuring that the construction would proceed with remarkable efficiency.

Assembly of the Metal Structure

Once the foundations were in place, the real challenge began: assembling the iron framework that would give the Eiffel Tower its iconic shape. This phase of construction spanned **21 months**, from **July 1887** to **March 1889**, and it involved extraordinary coordination between engineers, ironworkers, and laborers.

Phase 1: Initial Framework and the First Platform (July 1887 – March 1888)

The construction team began by erecting the four main pillars, or **legs**, of the tower. These legs were built at an angle and required intricate precision to ensure they met at the right points to form a solid base. Workers used **temporary scaffolding and cranes** mounted on rails to help position and raise the iron parts into place.

The initial challenge was to align the legs accurately, which had to meet at specific points to create the necessary structural balance. This alignment was achieved using **hydraulic jacks** that helped adjust the legs' positions to the required angles. The precision required in this phase was immense—any deviation could have caused instability in the entire structure.

Eiffel Tower: A Monumental Feat of Engineering

By **March 1888**, the team had completed the construction of the **first platform**, located at a height of about **57 meters** (187 feet). This marked a significant milestone, as it provided a stable base for the next phase of construction.

Phase 2: Second Platform and Height Growth (March 1888 – August 1888)

With the first platform in place, the workers continued assembling the iron structure upwards. The next major milestone was the completion of the **second platform**, located at **115 meters** (377 feet) above the ground.

During this phase, the engineers and workers faced increasing challenges. As the tower grew taller, the iron framework became more exposed to the elements, including **wind pressure**. The engineers had already accounted for these environmental factors in the tower's design, but the assembly required meticulous coordination to ensure the structure remained stable as it reached greater heights.

Despite these challenges, the **second platform** was completed in **August 1888**, just five months after the first. By this point, the Eiffel Tower had already become a towering presence on the Paris skyline, drawing widespread attention both locally and internationally.

Eiffel Tower: A Monumental Feat of Engineering

Phase 3: Reaching the Top and Final Assembly (August 1888 – March 1889)

The final stage of the Eiffel Tower's construction was perhaps the most dramatic. The workers continued assembling the structure, narrowing the tower as they ascended toward the **third platform**, which would reach a height of **276 meters** (905 feet). This part of the tower included the **cupola**, which would crown the structure.

The top section of the tower, which houses the **spire**, required particularly careful assembly, as it was both the most delicate and most visible part of the structure. Workers had to balance speed with precision to ensure the tower's aesthetic and structural integrity remained intact.

One of the most remarkable achievements during this phase was the use of **cranes and winches** that were mounted on the tower itself. These innovative machines allowed workers to continue raising the iron parts to unprecedented heights without relying solely on ground-based equipment.

By **March 15, 1889**, the final piece of the top section was installed, completing the Eiffel Tower's assembly. This milestone marked the culmination of nearly two years of tireless work and remarkable engineering.

Completion

Eiffel Tower: A Monumental Feat of Engineering

The Eiffel Tower was officially completed in **March 1889**, just in time for the **Exposition Universelle** (World's Fair), which opened on **May 6, 1889**. The tower served as the entrance arch to the fair and quickly became its most prominent attraction.

In total, the construction of the Eiffel Tower took **two years, two months, and five days** to complete—an extraordinary accomplishment considering the complexity and scale of the project. The tower's height and innovative design made it the tallest man-made structure in the world at the time, a title it held until the completion of the **Chrysler Building** in New York in 1930.

Challenges and Triumphs

While the timeline of the Eiffel Tower's construction may seem remarkably fast, it was not without its challenges. The sheer size of the project, combined with the need for precision in every aspect of the work, meant that any mistake could have resulted in significant delays or even structural failure. However, **Gustave Eiffel's leadership**, combined with the skill of the engineers and laborers, ensured that the project proceeded smoothly.

Precision and Innovation

One of the key reasons for the Eiffel Tower's success was the **precision of its design and assembly**. Every piece of

Eiffel Tower: A Monumental Feat of Engineering

iron used in the tower was prefabricated to exact specifications, allowing the workers to fit them together like a giant puzzle. The use of **rivets**, **hydraulic jacks**, and other innovations helped ensure that the structure was assembled quickly and securely.

Legacy of Time Management

The construction timeline of the Eiffel Tower is often cited as a model of efficiency and time management. In an era before modern construction equipment and techniques, Eiffel and his team demonstrated that careful planning, innovative design, and skilled labor could overcome even the most daunting challenges.

Conclusion

The Eiffel Tower's construction timeline is a remarkable story of speed, precision, and innovation. From the initial preparations in January 1887 to its completion in March 1889, the tower rose from the banks of the Seine to become one of the most iconic structures in the world. The efficiency with which it was built, combined with the challenges it overcame, makes the Eiffel Tower not only an architectural wonder but also a lasting testament to human ingenuity and perseverance.

Chapter 5: The Eiffel Tower's Legacy

Over 130 years since its completion, the **Eiffel Tower** has become a global icon, embodying the spirit of France and representing the zenith of 19th-century engineering. What was once a controversial structure that many Parisians wanted demolished has since evolved into one of the most visited monuments in the world. This chapter delves into the lasting legacy of the Eiffel Tower, its impact on architecture, engineering, communications, and tourism, and the cultural significance it has acquired over more than a century.

A Controversial Beginning

When the Eiffel Tower was first proposed by **Gustave Eiffel** for the **1889 World's Fair**, it was met with widespread skepticism and hostility. Many artists and intellectuals of the time, including prominent figures such as **Guy de Maupassant** and **Charles Gounod**, signed a petition against it, calling it a "useless and monstrous" structure that would ruin the beauty of Paris. Yet despite the outcry, Eiffel and his team persevered, constructing a tower that would not only stand the test of time but also become one of the most beloved landmarks in the world.

Eiffel Tower: A Monumental Feat of Engineering

The tower was initially intended to be a temporary structure, with a lifespan of just 20 years. Eiffel, however, recognized the potential for it to serve a practical purpose beyond the World's Fair. His foresight proved invaluable, as he convinced the French government to preserve the tower by demonstrating its usefulness as a **radio transmission tower**. This decision would secure the tower's place in history, allowing it to evolve from a controversial installation into a beloved symbol of technological progress.

A Symbol of Innovation and Industrial Might

The construction of the Eiffel Tower was a groundbreaking achievement in the late 19th century. It demonstrated the power of iron as a structural material and highlighted the advancements in industrial engineering that had emerged during the Industrial Revolution. For its time, the Eiffel Tower was the tallest structure in the world, standing at **300 meters (984 feet)** when it was completed. It retained this title until 1930, when it was surpassed by the **Chrysler Building** in New York.

One of the most remarkable aspects of the Eiffel Tower is how it was built—an astonishing feat of precision engineering. Using **7,300 tons of wrought iron, 2.5 million rivets**, and a team of **300 workers**, the tower rose from the ground over a period of just **21 months**. The

Eiffel Tower: A Monumental Feat of Engineering

methodical assembly process, with prefabricated iron pieces riveted together on-site, showcased new possibilities for mass-produced construction. The success of the Eiffel Tower paved the way for future advancements in skyscraper design and the widespread use of metal frameworks in architecture.

In its form, the Eiffel Tower was a bold departure from the traditional masonry structures that had dominated architectural design for centuries. Its open-lattice structure, which allowed wind to pass through the tower rather than pushing against a solid surface, made it both lighter and more resilient to environmental forces such as strong winds. This engineering ingenuity was not only a feat of the time but also laid the foundation for the development of modern skyscrapers.

Saving the Eiffel Tower: From Temporary to Permanent

Though it was initially planned as a temporary structure, Gustave Eiffel's brilliant use of the tower for **scientific experiments** and **telecommunications** helped ensure its survival. By 1909, when the tower's 20-year lease expired, it had already proven invaluable as a **radio transmission station**. Eiffel, always the visionary, had anticipated the rise of wireless technology and had lobbied for the tower to

Eiffel Tower: A Monumental Feat of Engineering
be used for experiments in telecommunications and weather monitoring.

In 1903, a wireless telegraphy station was installed at the tower, allowing for long-distance communications. By **1913**, the Eiffel Tower had successfully transmitted signals as far as **North America**, an extraordinary achievement for the time. Its strategic importance was further underscored during **World War I**, when the tower's radio station played a key role in intercepting enemy communications, including a message that led to the French victory in the **First Battle of the Marne**. This event cemented the Eiffel Tower's utility beyond its original purpose and guaranteed its continued existence.

The Eiffel Tower in the Modern Era

Today, the Eiffel Tower remains one of the most visited landmarks in the world, drawing nearly **7 million visitors annually**. Its continued popularity is a testament to its cultural significance, as well as the enduring fascination with its sheer scale and historical importance.

For tourists, the Eiffel Tower offers a stunning panorama of Paris, with three public viewing levels that allow visitors to experience the city from varying heights. The **third level**, at the very top of the tower, provides breathtaking views that stretch for miles on clear days. The tower is not only a tourist destination but also a symbol of **romance**,

Eiffel Tower: A Monumental Feat of Engineering **art**, and **culture**. It has inspired countless works of art, literature, and films, becoming a permanent fixture in the global imagination.

Over the decades, the Eiffel Tower has been the site of many significant events, celebrations, and even protests. It has served as a backdrop for fireworks displays, light shows, and special commemorations, such as the **2000 millennium celebration** when the tower was brilliantly illuminated. Its image is constantly evolving, with technological upgrades and maintenance ensuring its preservation for future generations. The tower has been repainted 19 times since its construction, with **60 tons of paint** required each time to prevent rust and maintain its iconic appearance.

A Masterpiece of Engineering and Cultural Symbolism

More than just a tourist attraction, the Eiffel Tower stands as a **cultural and engineering icon**. Its innovative design broke the mold of traditional architecture, offering a glimpse into the future of construction. Architects and engineers worldwide drew inspiration from the tower, which demonstrated that iron structures could be not only functional but also aesthetically pleasing.

For France, the Eiffel Tower represents national pride, innovation, and resilience. It embodies the industrial revolution and serves as a reminder of France's role as a

Eiffel Tower: A Monumental Feat of Engineering leader in technological progress during the late 19th century. Even though it was initially reviled by many, it has become a symbol of the very modernity and industrial prowess that critics once feared it would overshadow.

The Eiffel Tower and Technology

The Eiffel Tower continues to evolve with modern technology. It has been used as a **television transmission tower** since 1957, further cementing its role in the advancement of communications. In recent years, the tower has incorporated **LED lighting** and **renewable energy solutions** such as wind turbines and solar panels, demonstrating its adaptability to contemporary environmental concerns.

In addition to its role in broadcasting and telecommunications, the tower has been at the forefront of various scientific experiments, including studies related to **aerodynamics**, **meteorology**, and **astronomy**. Its height and structure have made it an ideal platform for these activities, allowing it to contribute to scientific knowledge even in the 21st century.

The Power of Teamwork

The construction of the Eiffel Tower stands as a lasting testament to the power of teamwork, coordination, and precision. The sheer scale of the project required intense

Eiffel Tower: A Monumental Feat of Engineering collaboration between engineers, architects, ironworkers, and laborers. Each piece of the tower was meticulously crafted and assembled, and the result was a structure that pushed the boundaries of what was thought possible at the time.

In many ways, the tower's legacy is not just about its physical form but about the people who built it. Gustave Eiffel's leadership, along with the hard work of thousands of individuals, demonstrated what could be achieved when a common goal was pursued with determination and ingenuity. The tower's construction and its continued survival serve as a reminder of the immense potential that lies in collaborative human endeavor.

Conclusion: An Enduring Legacy

The Eiffel Tower's legacy goes far beyond its function as a tourist attraction or radio transmission tower. It represents a unique combination of art, engineering, and cultural symbolism. What began as a temporary structure has become one of the most enduring icons in human history, a symbol of innovation, creativity, and the indomitable spirit of progress.

From its early days as a controversial addition to the Paris skyline to its modern status as a beloved global landmark, the Eiffel Tower continues to inspire awe and admiration. It stands as a testament not only to the genius of **Gustave**

Eiffel Tower: A Monumental Feat of Engineering
Eiffel and his team but also to the power of vision, perseverance, and the ability of humanity to create beauty out of industry. Over a century later, the Eiffel Tower remains a shining example of what can be achieved when creativity meets technology, and it will undoubtedly continue to captivate generations to come.

Conclusion

The construction of the **Eiffel Tower** was not just a technical achievement—it was a monumental project that pushed the boundaries of engineering, design, and human labor in the 19th century. At the time, nothing like it had ever been attempted. Its sheer scale, innovative use of materials, and the ambitious timeline made it one of the most extraordinary feats of the Industrial Revolution. When we reflect on the construction of the Eiffel Tower, we see more than just a towering iron structure. We see the culmination of vision, determination, and an unwavering belief in the possibilities of engineering and human collaboration.

Raw Materials: The Backbone of the Tower

The construction of the Eiffel Tower required an enormous amount of raw materials, most notably **wrought iron**—a material chosen for its durability and versatility. The fact that **7,300 tons** of this material were used to craft the elegant lattice design of the tower speaks to the scale of the project. In an era before the widespread use of steel, wrought iron was the material of choice for large-scale structures, and the Eiffel Tower demonstrated just how far it could be pushed. The iron used came from a single factory, reflecting the highly coordinated logistics of the

Eiffel Tower: A Monumental Feat of Engineering

time. Each piece was meticulously crafted and then transported to the construction site, where it was assembled like a massive jigsaw puzzle. The use of **2.5 million rivets**—heated, hammered, and fixed into place by skilled workers—was crucial to holding the iron beams together. Each rivet was essential, a tiny but critical part of the larger whole, ensuring the tower's stability and endurance through the decades.

The fact that the Eiffel Tower still stands strong today, more than a century after its construction, is a testament to the quality of the materials and the craftsmanship involved. The iron lattice framework, designed to minimize weight while maximizing strength, remains a brilliant example of architectural efficiency. The maintenance of the tower—such as the **60 tons of paint** regularly applied to prevent rust—also highlights the foresight of its builders. They understood that such a colossal structure would need ongoing care, ensuring its longevity for generations to come.

Manpower: The Human Effort Behind the Tower

While the materials were critical, the true heart of the Eiffel Tower's construction lies in the **300 skilled laborers** who brought it to life. These workers, many of them specialized ironworkers, were tasked with the formidable challenge of assembling the tower piece by piece. Working

Eiffel Tower: A Monumental Feat of Engineering
at heights that were, at the time, unprecedented for construction, they faced dangers that we can scarcely imagine today. Despite the lack of modern safety equipment, Gustave Eiffel was determined to minimize accidents, and his implementation of safety measures, including guardrails and protective screens, helped to keep fatalities remarkably low—an extraordinary achievement for such a risky endeavor.

The workers not only had to assemble the iron pieces but also ensure that each section was perfectly aligned. The Eiffel Tower's elegant tapering shape required precision at every step. Any mistake could have compromised the entire structure. Yet, through their hard work and expertise, they managed to erect the tower in just over **two years**, a timeline that was, and still is, incredibly ambitious for a project of this size and complexity. Their efforts were aided by Eiffel's decision to prefabricate many of the tower's components off-site. By assembling sections in advance, the on-site construction was significantly streamlined. This innovative approach to construction—using prefabricated parts—was a precursor to modern building techniques and demonstrated the efficiency of industrialized construction processes.

Engineering Precision: A Triumph of Innovation

Eiffel Tower: A Monumental Feat of Engineering

At the core of the Eiffel Tower's construction was a triumph of **engineering precision**. Gustave Eiffel, along with his team of engineers, including **Maurice Koechlin** and **Émile Nouguier**, pushed the limits of what was thought possible at the time. Their calculations had to account not only for the sheer weight of the tower but also for environmental factors, such as the powerful winds that would batter the structure at such great heights. The innovative use of a **lattice design**—wherein the iron beams crisscrossed to create a framework that was both lightweight and sturdy—was critical to the tower's success. This design allowed the Eiffel Tower to stand tall without collapsing under its own weight, and it ensured that the structure could withstand the forces of nature, swaying slightly but remaining secure in even the strongest gusts.

The Eiffel Tower also marked a significant departure from traditional architectural styles. In the 19th century, most monumental buildings were constructed using heavy masonry. By contrast, the Eiffel Tower's iron structure was light, open, and skeletal, a design that many initially criticized as ugly or too industrial. However, this same design is what has allowed the tower to endure for so long, without the need for extensive structural reinforcement. In fact, the Eiffel Tower was so ahead of its time that it paved the way for future skyscrapers and tall structures, demonstrating that height could be achieved through careful engineering rather than sheer mass.

Eiffel Tower: A Monumental Feat of Engineering

Eiffel's understanding of physics and material science played a crucial role in the tower's design. He was aware of the wind loads the structure would face and designed the tower to resist them. The open lattice structure minimized wind resistance, while the tower's tapering shape distributed weight evenly and allowed it to withstand significant environmental stresses. The fact that the tower has survived the test of time, including World War II, further underscores the brilliance of its design and construction.

The Tower as a Monument to Human Achievement

Beyond its raw materials, manpower, and engineering marvels, the Eiffel Tower stands as a **monument to human achievement**. It is not just a symbol of Paris or France; it is a symbol of what humans can accomplish when they harness creativity, science, and hard work. At the time of its construction, the tower represented the pinnacle of industrial achievement. It was the tallest man-made structure in the world, standing at a height of 300 meters (eventually reaching 324 meters with the addition of antennas). It was a bold statement that technology and industry had come of age.

For the people who built it, the Eiffel Tower was more than just a job. It was a project that pushed them to their limits, requiring them to innovate, problem-solve, and work

Eiffel Tower: A Monumental Feat of Engineering

together. Today, it serves as a reminder of their efforts—a lasting tribute to their skill, perseverance, and ingenuity. The tower's survival and continued relevance over 130 years later is a testament to its creators' foresight. Initially intended as a temporary structure for the **1889 World's Fair**, the Eiffel Tower was supposed to be dismantled after 20 years. However, its utility as a radio transmission tower and its growing popularity as a tourist attraction saved it from destruction.

An Enduring Legacy

As we stand in awe of the Eiffel Tower today, it's easy to take its existence for granted. But it's important to remember the enormous effort, planning, and ingenuity that went into its creation. From the 7,300 tons of iron that form its skeletal frame to the millions of rivets that hold it together, the tower is a testament to what can be achieved when human ambition and engineering expertise come together. The Eiffel Tower is not just an iron structure—it is a symbol of possibility. It represents the spirit of invention, the ability to dream big, and the determination to bring those dreams to life. Its legacy continues to inspire engineers, architects, and dreamers around the world.

In conclusion, the Eiffel Tower stands tall not just as a physical structure but as a **monument to human ingenuity**, collaboration, and perseverance. It is a reminder

Eiffel Tower: A Monumental Feat of Engineering that with the right combination of vision, materials, labor, and engineering, even the most ambitious goals can be realized. The Eiffel Tower may have been built over two years, but its impact will be felt for centuries to come—a lasting tribute to human achievement and a beacon of innovation for future generations.

Bibliography

1. **"The Eiffel Tower"** by Bertrand Lemoine

A comprehensive history of the Eiffel Tower, covering its construction, its initial reception, and its evolution as a cultural icon.

2. **"Eiffel's Tower: The Thrilling Story Behind Paris's Beloved Monument and the Extraordinary World's Fair That Introduced It"** by Jill Jonnes

This book intertwines the story of the Eiffel Tower's construction with the broader context of the 1889 World's Fair.

3. **"Gustave Eiffel: The Man Behind the Masterpiece"** by David I. Harvie

A biography of Gustave Eiffel that delves into his life, career, and the projects that made him a legend in the engineering world.

4. **"The History of Modern Architecture"** by Leonardo Benevolo

A global overview of architectural innovation, with discussions on how structures like the Eiffel Tower influenced modern design.

Eiffel Tower: A Monumental Feat of Engineering

5. **"Paris Reborn: Napoleon III, Baron Haussmann, and the Quest to Build a Modern City"** by Stephane Kirkland

Although not exclusively about the Eiffel Tower, this book explores the modernization of Paris, which set the stage for the tower's construction.

6. **"Structural Iron and Steel, 1850–1900"** by Robert Thorne

A detailed look at the use of iron and steel in construction during the late 19th century.

7. **"Paris: The Biography of a City"** by Colin Jones

A historical exploration of Paris from ancient times to the modern day, including the construction of the Eiffel Tower.

Eiffel Tower: A Monumental Feat of Engineering

Acknowledgments

Writing *The Eiffel Tower: A Monumental Feat of Engineering* has been an incredibly rewarding journey, and I could not have completed this book without the support and guidance of many individuals.

First and foremost, I would like to express my deepest gratitude to the brilliant minds behind the Eiffel Tower, particularly **Gustave Eiffel**, **Maurice Koechlin**, and **Émile Nouguier**, whose groundbreaking vision and engineering mastery continue to inspire generations. Their work has provided not only the foundation of this book but also a timeless example of human achievement.

To the countless historians, architects, and engineers whose research and documentation of the Eiffel Tower made this book possible, I owe immense thanks. The wealth of knowledge they have contributed to the world has allowed me to dive deep into the technical details, ensuring that this book remains informative and comprehensive.

A special thank you to my family and friends for their unwavering encouragement and support throughout the writing process. Your belief in this project kept me motivated and focused.

Eiffel Tower: A Monumental Feat of Engineering

I would also like to extend my appreciation to my editor, whose keen eye and invaluable insights helped shape this book into what it is today. Your feedback and dedication have been instrumental in bringing clarity and precision to the narrative.

Lastly, I want to thank **you**, the reader, for your curiosity and passion for engineering marvels like the Eiffel Tower. I hope this book deepens your appreciation for the incredible feat that it represents and inspires you to explore the remarkable possibilities of human ingenuity.

This book is as much a tribute to those who built the Eiffel Tower as it is to those who continue to marvel at its beauty and brilliance. Thank you all for being a part of this journey.

Sincerely,

Zahid Ameer
Versatile Indie Author

Disclaimer

The information provided in this book, *The Eiffel Tower: A Monumental Feat of Engineering*, is based on historical records, research, and publicly available data about the construction of the Eiffel Tower. Every effort has been made to ensure the accuracy of the facts presented, but the author and publisher make no representations or warranties regarding the completeness or current accuracy of the information. This book is intended for informational and educational purposes only and should not be considered a definitive historical account.

The author and publisher disclaim any liability for any errors or omissions or any adverse consequences arising from the use or interpretation of the material contained in this book. Readers are encouraged to conduct their own research or consult additional sources to verify the details discussed herein. All opinions expressed in this book are those of the author and do not necessarily reflect the views of any institutions or organizations associated with the Eiffel Tower.

About me

I am Zahid Ameer, hailing from the vibrant country of India. As an author, ghostwriter, bibliophile, online affiliate marketer, blogger, YouTuber, graphic designer, and animal lover, I have woven my passions into a unique tapestry that defines my life's work.

Born and raised in India, I have always possessed a deep love for literature. With an insatiable appetite for books, I have amassed an impressive collection of around 1,600 titles, predominantly in English. My passion for reading brings me immense joy and serves as a source of inspiration for my writing endeavors.

I have compiled an impressive portfolio of written works as an author and ghostwriter. With a captivating writing style and an innate ability to craft engaging narratives, I bring my stories to life, captivating readers from all walks of life. My wide range of interests and experiences contribute to the richness of my writing, allowing me to connect with my audience on a heartfelt level effortlessly.

Beyond my literary pursuits, I have also established a strong presence on various digital platforms. I utilize my YouTube channel and blog to raise awareness about all types of knowledge and to share heartwarming stories of animals. Using my platform to shed light on important

Eiffel Tower: A Monumental Feat of Engineering issues, I strive to create a world where humans and animals can coexist harmoniously.

In addition to my work as an author, I have also dabbled in the world of affiliate marketing. With my webpreneur spirit, I have ventured into online marketing, leveraging my knowledge and skills to promote products and services that align with my values.

However, my most cherished role is that of a father. Family is at the core of my being, and everything I do is centered around creating a better future for my loved ones. My dedication to my family is evident in my passion for personal growth and my relentless pursuit of success. Through my various endeavors, I strive to set an example of perseverance and ambition for my children, inspiring them to chase their dreams unapologetically.

In a world where specialization often dominates, I defy convention by embracing multiple passions and excelling in diverse fields. My love for books, animals, and family has become the driving force behind my achievements. By the grace of Almighty God, my unique blend of characteristics has allowed me to leave an indelible mark on the world, enriching the lives of those I encounter along the way.

To your grand success in life,

Eiffel Tower: A Monumental Feat of Engineering

Zahid Ameer
Versatile Indie Author

www.ingramcontent.com/pod-product-compliance
Lightning Source LLC
Chambersburg PA
CBHW070128230526
45472CB00004B/1467